# Section 7

## Genetics, populations, evolution and ecosystems

The basis of the theory of evolution is that all new species arise from existing species. Therefore species share a common ancestry, as reflected in phylogenetic classification. There is a great deal of molecular and biochemical evidence to support the theory of evolution. The members of a species share the same genes but not necessarily the same alleles.

Species exist in different populations that show variation in their phenotypes because of genetic and environmental factors. Allele frequencies within these populations can change as a result of genetic drift and natural selection. New species arise when these populations become reproductively isolated.

Communities are made up of different populations. Populations are affected by biotic and abiotic factors.

### Inheritance

The genetic make-up of an organism is its genotype. The organism's phenotype results from the interaction of its genotype with the environment. Genes may have several variations, or alleles. These may be recessive, dominant or codominant. Diploid organisms have two copies of each gene, one on each homologous chromosome. The alleles may be homozygous or heterozygous.

Genetic diagrams can be used to explain or predict the results of monohybrid and dihybrid crosses. Monohybrid crosses involve one gene with two or more alleles. Dihybrid crosses involve two genes. There are also sex-linked genes. Some genes have multiple alleles. Epistasis occurs when several genes control one characteristic. Autosomal linkage occurs between genes on the same chromosome.

**1 Explain the difference between an organism's genotype and its phenotype. (AO1)** 2 marks

..........................................................................................

..........................................................................................

..........................................................................................

..........................................................................................

**2 Explain the difference between a gene and an allele. (AO1)** 2 marks

..........................................................................................

..........................................................................................

..........................................................................................

..........................................................................................

**3 Complete the table by adding the definitions of the terms. (AO1)** 4 marks

| Term | Definition |
|---|---|
| Homozygous | |
| Heterozygous | |
| Dominant | |
| Recessive | |

**4** Manx cats have no tail as the result of a dominant allele, T. Cats with normal tails have the genotype tt. Embryos with the genotype TT do not survive.

Complete a genetic diagram to show the likely ratio of genotypes and phenotypes from a cross between two heterozygous Manx cats. (AO2) **4 marks**

*Parent phenotypes* Manx Manx

*Parent genotypes* .................... ....................

*Gametes* .................... ....................

*Offspring genotypes* .................... ....................

*Offspring phenotypes and ratio* .................... ....................

**5** Cystic fibrosis is caused by the recessive allele, f. The normal allele is dominant. Two parents who do not have cystic fibrosis have a baby with cystic fibrosis. Complete a genetic diagram to show the parents the probability of their next child also having cystic fibrosis. (AO2) **4 marks**

*Parent phenotypes* Normal Normal

*Parent genotypes* .................... ....................

*Gametes* .................... ....................

*Offspring genotypes* .................... ....................

*Offspring phenotypes* .................... ....................

*Probability of the next child having cystic fibrosis* ....................

**6** The pedigree shows the inheritance of a characteristic in one family.

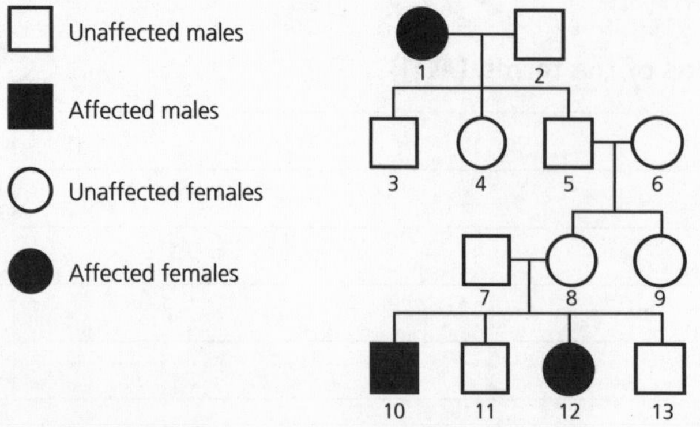

Unaffected males

Affected males

Unaffected females

Affected females

1 2

3 4 5 6

7 8 9

10 11 12 13

# AQA A LEVEL YEAR 2

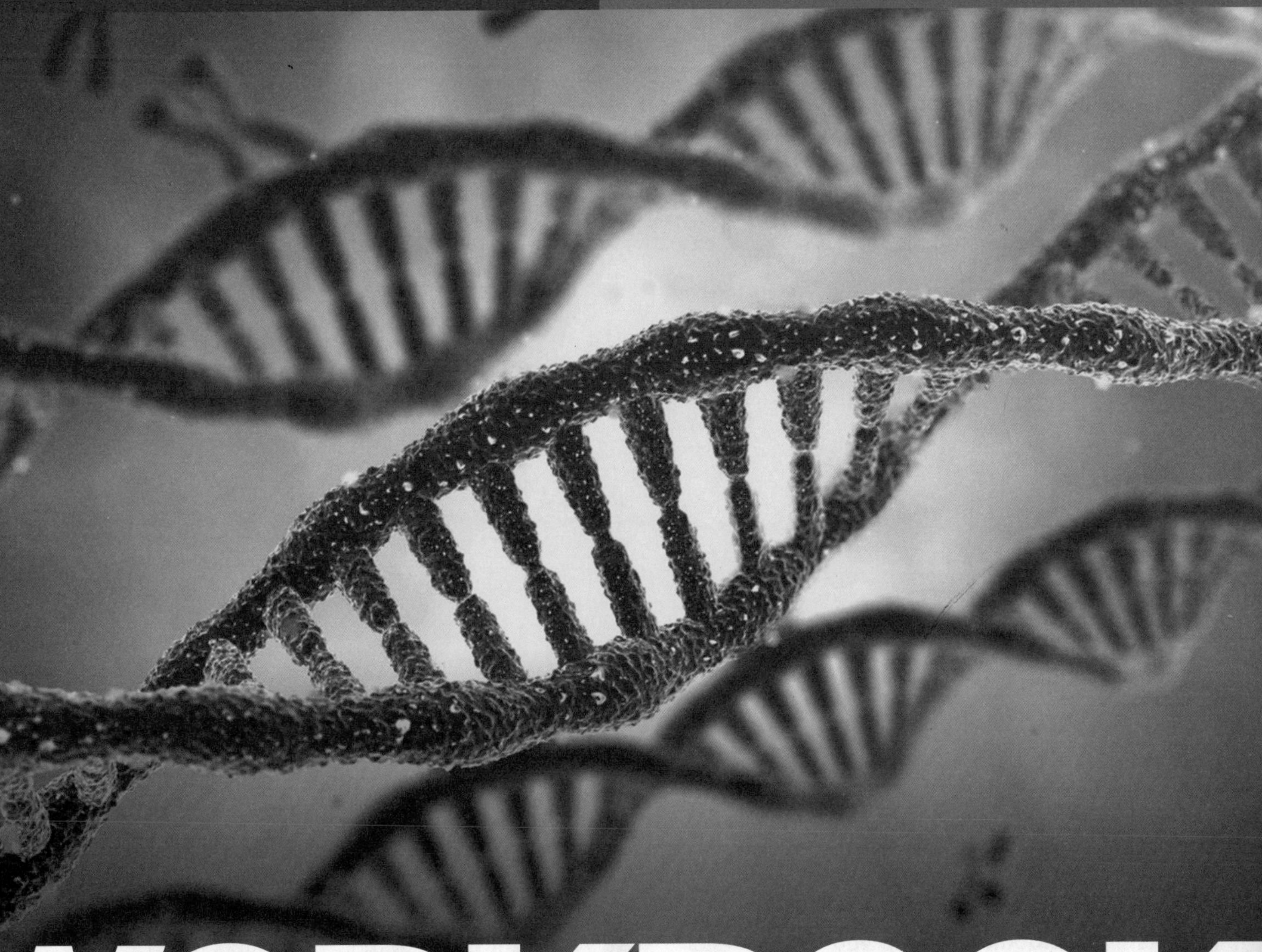

# WORKBOOK

## Biology

FOR THE 2015 SPECIFICATIONS

Genetics, populations, evolution and ecosystems • The control of gene expression

Pauline Lowrie

# Contents

WORKBOOK



1. **This workbook will help you** to prepare for your AQA A-level exam.
2. **Your exam** is 2 hours long for each of three A-level papers.
3. **For each topic** there are:
   - stimulus materials, including key terms and concepts
   - short-answer questions that build up to exam-style questions
   - spaces for you to write or plan your answers
   - questions that test your mathematical skills
4. **Answering the questions** will help you to build your skills and meet the assessment objectives AO1 (knowledge and understanding), AO2 (application) and AO3 (analysis).
5. **You still need to** read your textbook and refer to your revision guides and lesson notes.
6. **Mark schemes** are provided for some of the questions so that you can see what the examiner wants and where the marks come from.
7. **Timings** are given for the exam-style questions to make your practice as realistic as possible.
8. **Answers** are available at: www.hoddereducation.co.uk/workbookanswers

**a** **Is this characteristic controlled by a dominant or a recessive allele? Give evidence to support your answer. (AO2)** 1 mark

..............................................................................................................................

..............................................................................................................................

**b** **Give the possible genotype(s) of the following individuals: (AO2)** 3 marks

**3** ..............................................................................................................................

**7** ..............................................................................................................................

**11** ..............................................................................................................................

**7** **Red coat colour in cattle is determined by the allele $C^R$, which is codominant with the allele for white coat colour, $C^W$. Heterozygous cattle have a mixture of red and white hairs, a coloration that is called roan.**

**Another allele, H, results in hornless cattle. The recessive allele, h, results in horned cattle.**

**Complete a genetic diagram to show the possible phenotypes in the offspring of two roan cattle, heterozygous for the hornless allele, and the ratio in which they would be expected to be produced. (AO2)** 4 marks

***Parent phenotypes*** .............................. ..............................

***Parent genotypes*** .............................. ..............................

***Gametes*** .............................. ..............................

***Offspring genotypes*** .............................. ..............................

***Offspring phenotypes and ratios*** .............................. ..............................

**8** **ABO blood groups are determined by three alleles. $I^A$ and $I^B$ are codominant, while $I^O$ is recessive. The four blood groups and their genotypes are shown in the table.**

| Blood group | Possible genotypes |
|---|---|
| A | $I^AI^A$ or $I^AI^O$ |
| B | $I^BI^B$ or $I^BI^O$ |
| AB | $I^AI^B$ |
| O | $I^OI^O$ |

In addition, Rhesus blood groups are determined by a different allele. People with the genotype DD or Dd are Rhesus positive. People with the genotype dd are Rhesus negative.

A man who is blood group A, Rhesus positive, and a woman who is blood group B, Rhesus negative, have a child who is group O, Rhesus negative. Complete a genetic diagram to show the possible genotypes and phenotypes that their next child might have. (AO2)

4 marks

| | | |
|---|---|---|
| *Parent phenotypes* | Group A, Rhesus positive | Group B, Rhesus negative |
| *Parent genotypes* | ........................ | ........................ |
| *Gametes* | ........................ | ........................ |

| | | |
|---|---|---|
| Offspring genotypes | ........................ | ........................ |
| Offspring phenotypes and ratios | ........................ | ........................ |

**9** Two alleles on the X chromosome determine coat colour in cats. $X^G$ codes for ginger fur and $X^B$ codes for black fur. Female cats that are heterozygous have patches of black and ginger fur, which is called tortoiseshell.

Complete a genetic diagram to show the possible offspring of a tortoiseshell female and a black male cat. (AO2)

4 marks

| | | |
|---|---|---|
| *Parent phenotypes* | Tortoiseshell female | Black male |
| *Parent genotypes* | ........................ | ........................ |
| *Gametes* | ........................ | ........................ |

| | | |
|---|---|---|
| *Offspring genotypes* | ........................ | ........................ |
| *Offspring phenotypes* | ........................ | ........................ |

**10** **In a particular kind of flower, two genes determine flower colour, as shown below:**

Colourless precursor —(Enzyme A coded for by allele **A**)→ Colourless precursor 2 —(Enzyme B coded for by allele **B**)→ Purple pigment

a **Explain why a flower with the genotype aaBB has white flowers. (AO1)** 3 marks

........................................................................................................

........................................................................................................

........................................................................................................

........................................................................................................

........................................................................................................

........................................................................................................

b **Give the phenotypes of the following flowers. (AO1)** 3 marks

| Genotype | Phenotype |
|---|---|
| AaBb | |
| AAbb | |
| aaBB | |

**11** **In Labrador dogs, two genes control coat colour, as shown below:**

| Genotype | Phenotype |
|---|---|
| **B_E_** | Black |
| **bbE_** | Brown |
| **B_ee** | Golden |
| **bbee** | Golden |

**Use a genetic diagram to show the results of a cross between two Labradors of genotype BbEe. (AO2)** 4 marks

***Parent phenotypes*** .............................. ..............................

***Parent genotypes*** **BbEe** **BbEe**

***Gametes*** .............................. ..............................

***Offspring genotypes*** .............................. ..............................

***Offspring phenotypes*** .............................. ..............................

**12** **In an investigation, scientists crossed two plants that were both heterozygous for purple flowers and long pollen (PpRr). The following results were obtained:**

| Offspring phenotype | Observed number | Ratio | Expected number | Ratio |
|---|---|---|---|---|
| Purple flowers, long pollen | 296 | 15.6 | 240 | 9 |
| Purple flowers, round pollen | 19 | 1.0 | 80 | 3 |
| Red flowers, long pollen | 27 | 1.4 | 80 | 3 |
| Red flowers, round pollen | 85 | 4.5 | 27 | 1 |

**a The scientists decided to carry out a chi-squared test on these results. What was their null hypothesis? (AO2)** 1 mark

..........

..........

**b i How many degrees of freedom were there in this calculation? (AO1)** 1 mark

..........

**ii They obtained a value of $p > 0.05$. What does this mean? Use the words 'probability' and 'chance' in your answer. (AO2)** 2 marks

..........

..........

..........

..........

**c Suggest an explanation for these results being very different from the expected 9:3:3:1 ratio. (AO2)** 3 marks

..........

..........

..........

..........

..........

..........

**13** **Two pea plants heterozygous for height and seed colour were crossed. The following results were obtained:**

| Phenotype | Number | Expected number |
|---|---|---|
| Tall, green seeds | 80 | |
| Tall, yellow seeds | 40 | |
| Short, green seeds | 32 | |
| Short, yellow seeds | 8 | |

**a The scientists thought that these results were approximately a 9:3:3:1 ratio. Complete the table to show the expected numbers for a 9:3:3:1 ratio. (AO2)** 2 marks

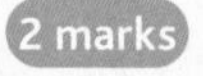

b The critical value of chi-squared for $p < 0.05$ is 7.82. The scientists' calculated value of chi-squared was lower than this value. What does this mean? (AO2) 2 marks

.................................................................................................................................................

.................................................................................................................................................

.................................................................................................................................................

.................................................................................................................................................

11

## Exam-style questions

**1** **The fruit fly, *Drosophila*, has four different colours of pigment in the eye. These pigments are produced in a metabolic pathway involving three enzymes, A, B and C. Each enzyme is coded for by two alleles, A and a, B and b, C and c respectively. In each case, the dominant allele codes for a functional enzyme and the recessive allele for a non-functional enzyme.**

Enzyme A, Enzyme B, Enzyme C

Vermillion pigment → Cinnabar pigment → Brown pigment → Red pigment

a **Name the kind of gene interaction involved in this process.** 1 mark

.................................................................................................................................................

b **Give the phenotypes of each of the following flies.** 3 marks

| Genotype | Phenotype |
|---|---|
| AabbCC | |
| aaBBCC | |
| AABbCc | |

**In *Drosophila*, males are XY and females XX. There is a sex-linked recessive allele for goggle eyes, which are unusually prominent eyes. Complete the genetic diagram to show the results of a cross between a female heterozygous for goggle eyes and a normal male.** 4 marks

***Parent phenotypes*** .................................... ....................................

***Parent genotypes*** .................................... ....................................

***Gametes*** .................................... ....................................

***Genotypes of offspring*** ............................................................................

***Ratio of phenotypes in offspring*** ..................................................................

**2** **The pedigree diagram shows the inheritance of achondroplasia in one human family.**

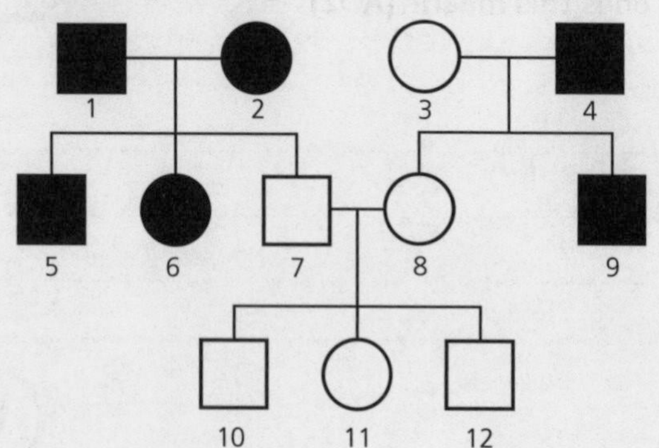


Squares = males

Circles = females

Shaded shapes = individuals with achondroplasia

**a** **Give the evidence from the diagram that achondroplasia is determined by a dominant allele.** 
1 mark

..........................................................................................................

..........................................................................................................

**b** **Fruit flies (*Drosophila*) normally have bright red eyes. This colour pigment results from a mixture of an orange pigment produced by one enzyme and a brown pigment produced by a different enzyme, as shown in the diagram.**

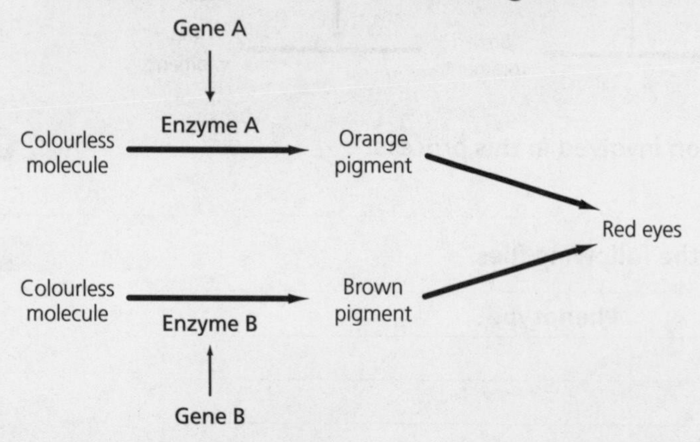


**The enzymes are coded for by two different genes. In each case, the dominant allele produces the functional enzyme and the recessive allele produces a non-functional enzyme.**

**Complete the diagram to show the results of a cross between two flies of genotype AaBb.** 
4 marks

***Parent phenotypes*** .................................... x ....................................

***Parent genotypes*** AaBb x AaBb

***Gametes*** .................................... x ....................................

***Genotypes of offspring:*** ..........................................................................

***Ratio of phenotypes in offspring*** ..........................................................................

## Populations

A population is all the organisms of one species living in a particular area at the same time. The gene pool is made up of all the alleles in a population. The frequency of alleles, genotypes and phenotypes within a population can be calculated using the Hardy-Weinberg formula. For the Hardy-Weinberg formula to predict these frequencies accurately, certain conditions must prevail.

**1** **Phenylketonuria (PKU) is a metabolic disorder that occurs in about 1 in 10000 babies born in the UK. It is caused by a recessive allele. Use the Hardy-Weinberg formula to calculate the frequency of heterozygotes in the population. Show your working. (AO2)** 3 marks

**2** **Give *four* assumptions that are made when using the Hardy-Weinberg equation. (AO1)** 4 marks

**3** **In humans, unattached earlobes are dominant over attached earlobes. In a certain population, 64% of people have unattached earlobes. What percentage of people are heterozygous? Show your working. (AO2)** 3 marks

## Evolution may lead to speciation

Individuals within a population show variation as a result of both genetic and environmental factors. Genetic variation results from mutation, meiosis and random fertilisation, which produce new combinations of alleles.

Natural selection means that organisms with phenotypes that give them a survival advantage are more likely to reproduce and so pass on their favourable alleles to their offspring. This leads to changes in the allele frequencies in the gene pool. Three types of selection operating on populations are known as stabilising, directional and disruptive. Evolution is a change in the frequency of alleles in a population. When the genetic differences result in members of one population being unable to interbreed with members of another population to produce fertile offspring, a new species has formed. Speciation may be sympatric or allopatric. In small populations, genetic drift can cause significant changes in allele frequencies.

**1** **Give *two* ways in which meiosis contributes to variation. (AO1)** 2 marks

........................................................................................................................

........................................................................................................................

**2** **Apart from the answers that you have given in question 1, give *two* reasons why the phenotypes in a population show genetic variation. (AO1)** 2 marks

........................................................................................................................

........................................................................................................................

........................................................................................................................

........................................................................................................................

**3** **The main reason why poachers shoot and kill elephants is to obtain the tusks, which are made of ivory. Researchers at a national park in Uganda found that 15% of female elephants and 9% of male elephants are born without tusks. In 1930 the figure was only 1%. At another national park in Zambia, 38% of elephants have no tusks.**

**a** **Use your knowledge of natural selection to explain these figures. (AO2)** 3 marks

........................................................................................................................

........................................................................................................................

........................................................................................................................

........................................................................................................................

........................................................................................................................

........................................................................................................................

**b** **What kind of selection is shown by these elephants? Explain your answer. (AO1, AO2)** 2 marks

........................................................................................................................

........................................................................................................................

**4** **In 1982 and 1983 nearly half the cheetahs in a wildlife park in the USA died from a disease epidemic. The same disease rarely kills other kinds of cat. Cheetahs are all genetically similar. Scientists think that there was an event about 10000 years ago that killed large numbers of cheetahs.**

**Use the information in the passage and your own knowledge to suggest why the disease killed so many cheetahs. (AO2)** 3 marks

5 Jaguars live in a varied habitat in South America, ranging from grassland to dense forests. Jaguars often have an orange colour with black spots, while others are black. Scientists have suggested that spotted jaguars are camouflaged in grassland and areas with trees, while black jaguars are adapted for hunting at night in dense forests. However, jaguars with intermediate coat colours are not found. Explain how this is an example of disruptive selection. (AO2) **2 marks**

6 Clutch size (the number of eggs laid by a female bird in one season) is genetically determined. Scientists studying a species of bird found that most birds of that species laid five eggs in a clutch. Very few birds had clutch sizes much greater or smaller than this. Use your knowledge of selection to explain why. (AO2) **3 marks**

7 Explain why genetic drift is important only in small populations. (AO1) **2 marks**

8 There are two separate species of squirrel around the Grand Canyon in Arizona. One species is on the north side and the other species is on the south side.

a How could a scientist show that these are two separate species? (AO1) **2 marks**

**b** **Suggest how these two species may have evolved from a single population that was present in the area before the Grand Canyon formed. (AO2)** 5 marks

........................................................................................................

........................................................................................................

........................................................................................................

........................................................................................................

........................................................................................................

........................................................................................................

........................................................................................................

........................................................................................................

........................................................................................................

........................................................................................................

**9** **In North America there is a type of fly that lays its eggs in hawthorn fruits. When the apple was introduced to North America, some of these flies laid their eggs in apple fruits instead. There are now two types of fly and scientists believe that they will soon be different species.**

**a** **What change would be needed before scientists can call these two kinds of fly different species? (AO1)** 2 marks

........................................................................................................

........................................................................................................

........................................................................................................

........................................................................................................

**b** **What kind of speciation may be taking place? Explain your answer. (AO1)** 2 marks

........................................................................................................

........................................................................................................

........................................................................................................

........................................................................................................

13

## Exam-style questions

**1** **There are about 400 Californian condors in the world, living at a handful of sites in central and North America. In 1987 the wild population fell to just 22. These remaining birds were caught for captive breeding. All the wild condors in the world today can be traced back to 16 of these birds.**

**About 10% of the condors carry a recessive allele for a lethal form of dwarfism. Chicks that inherit this condition do not survive to adulthood.**

**a** **Alleles that code for disadvantageous conditions such as a lethal form of dwarfism are usually very rare in a population. Use your knowledge of natural selection to explain why.** 3 marks

.......................................................................................................

.......................................................................................................

.......................................................................................................

.......................................................................................................

.......................................................................................................

.......................................................................................................

.......................................................................................................

**b** **Use the information to explain why this allele is relatively common in the wild condor population.** 3 marks

.......................................................................................................

.......................................................................................................

.......................................................................................................

.......................................................................................................

.......................................................................................................

.......................................................................................................

**2** **In 1708, a small number of German Baptists migrated to the USA and formed a group in Pennsylvania. They are a fairly strict religious group who usually marry within their community. In 1950, there were about 3500 members of this community. A geneticist studied the incidence of MN blood groups in 200 members of this population. This blood group is controlled by a single gene that has two alleles, as follows:**

| Blood group | Genotype |
|---|---|
| M | Homozygous for **M** allele |
| N | Homozygous for **N** allele |
| MN | Heterozygous |

**a** **The geneticist found that 102 individuals had blood group M, 96 had MN, and 2 had type N. Calculate the frequencies of the alleles. Show your working.**

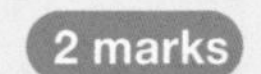


**The table shows the frequency of these blood groups in modern Germany and in the USA.**

| | Frequency of blood group/% | | |
|---|---|---|---|
| | M | N | MN |
| Germany | 30 | 20 | 50 |
| USA | 29 | 21 | 50 |

**b Suggest an explanation for the frequency of these blood groups in the German Baptist community being very different from these figures.** 3 marks

.......................................................................................................................

.......................................................................................................................

.......................................................................................................................

.......................................................................................................................

.......................................................................................................................

.......................................................................................................................

## Populations in ecosystems

A community is made up of many different populations. The ecosystem is the community and its interactions with the non-living components of the environment. Species occupy a specific niche within their habitat. The carrying capacity is the size of population that an ecosystem is able to support. This can vary, dependent upon a range of biotic and abiotic factors.

There are different techniques used to estimate population size. Which technique is used will depend on whether the organisms are motile or non-motile.

Succession is the changes in the species present in an environment over a period of time. At each stage of succession, species change the environment, making it more suitable for other organisms to grow there.

**1 Match the terms in the table to their definitions. (AO1)** 4 marks

| Term | Definition |
|---|---|
| 1 Niche | A All the members of one species that live in the same area at the same time |
| 2 Community | B All the living organisms in an area and their interactions with the non-living components of the environment |
| 3 Ecosystem | C The place where an organism lives and the role that an organism plays within the ecosystem |
| 4 Population | D All the members of all the species in the same place at the same time |

**2 What is meant by the *carrying capacity* of a population? (AO1)** 2 marks

.......................................................................................................................

.......................................................................................................................

.......................................................................................................................

.......................................................................................................................

**3 Isle Royale is an island in North America that is uninhabited by humans. Moose were introduced to the island in about 1908. Their population size increased as they ate the vegetation. Visitors to the island in 1934 found many dead moose and the others were weak and thin. There was very little vegetation left on the island.**

**In 1948, wolves walked across ice to the island. They started to kill the moose for food. The island now has about 1000 moose and 24 wolves. The populations remain relatively stable in numbers, and the vegetation seems to be in good supply for the moose to eat.**

a Explain why the moose population increased rapidly when they first populated the island. (AO2) **2 marks**

b What kind of competition did the moose experience in 1934 and why? (AO1) **2 marks**

c Explain why the moose and wolf populations are now relatively stable in numbers. (AO2) **4 marks**

d Use this example to explain what is meant by 'carrying capacity'. (AO1) **2 marks**

**4** After a forest fire has destroyed the vegetation in an area, succession occurs. Describe this process. (AO1) **6 marks**

**5** **The South Downs is an area of chalk grassland that has been grazed by animals for centuries. There are many species found only in this area.**

**a** **Suggest and explain what might happen to the grassland if sheep were removed. (AO1)** **3 marks**

..............................................................................................................................

..............................................................................................................................

..............................................................................................................................

..............................................................................................................................

..............................................................................................................................

..............................................................................................................................

**b** **Conservationists ensure that grazing continues so that the grassland habitat remains. Suggest why. (AO2)** **3 marks**

..............................................................................................................................

..............................................................................................................................

..............................................................................................................................

..............................................................................................................................

..............................................................................................................................

..............................................................................................................................

**6** **Explain why a farmer's field does not undergo the process of succession. (AO2)** **2 marks**

..............................................................................................................................

..............................................................................................................................

**7** **The graph shows the population of sheep on an island over 100 years.**

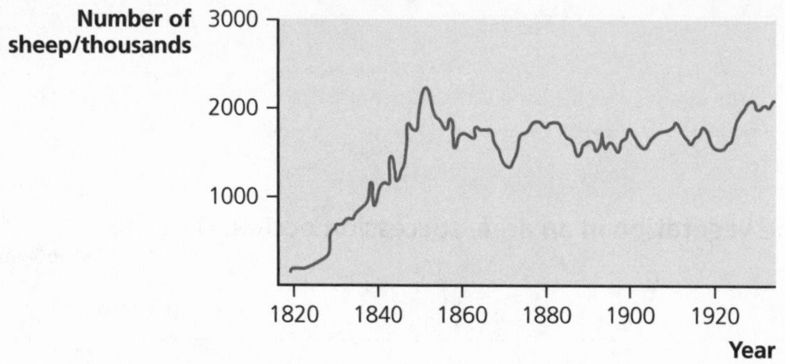

Number of sheep/thousands: 1000, 2000, 3000

Year: 1820, 1840, 1860, 1880, 1900, 1920

**a** **Sheep were introduced to the island in 1800. Explain why the population grew rapidly over the next 30 years. (AO2)** **2 marks**

..............................................................................................................................

..............................................................................................................................

..............................................................................................................................

..............................................................................................................................

**b i Use the graph to estimate the carrying capacity of the sheep. (AO1)** 1 mark

..........

**ii The sheep population fell between 1850 and 1860. Suggest an explanation. (AO2)** 2 marks

..........

..........

..........

..........

**8 The capture-mark-recapture method was used to estimate a population of seals. Sixty seals were caught, tagged and released. A few days later, 120 seals were caught at random. Thirty of these 120 seals were tagged. Calculate the size of the seal population. Show your working. (AO2)** 2 marks

**9 Give *two* assumptions that are made when using the capture-mark-recapture method of estimating population size. (AO1)** 2 marks

..........

..........

..........

..........

**10 A student was asked to use the capture-mark-recapture method to estimate the population of harvest mice in a meadow. Give *two* factors that she would need to consider when choosing a suitable method of marking the harvest mice. (AO3)** 2 marks

..........

..........

..........

..........

**11** **The lion population in Zimbabwe has fallen sharply over recent decades because of hunting. Some people will pay a considerable sum of money to shoot a male lion as a 'trophy'. The money that they pay makes an enormous difference to the lives of people living in Zimbabwe.**

**A complete ban on hunting was imposed in Zimbabwe, and the lion population has increased again. A conservation organisation made a proposal that they should allow licensed hunting of a small number of male lions each year, both to generate money to pay for conservation, and to improve the standard of living of the local people. Evaluate this proposal. (AO3)** 4 marks

.................................................................................................

.................................................................................................

.................................................................................................

.................................................................................................

.................................................................................................

.................................................................................................

.................................................................................................

.................................................................................................

## Required practical 12: Investigation into the effect of a named environmental factor on the distribution of a given species

**A student carried out an investigation to record the distribution of plant species along a transect through sand dunes. She laid a measuring tape along the sand dunes, from the front of the dunes to woodland at the back. She placed a 1 $m^2$ quadrat against the measuring tape at 5 m intervals along the tape line. In each quadrat she recorded the number of plants. She also took a soil sample every 25 m so that she could measure the soil moisture, the organic content and the pH. The tables show some of her results.**

| Distance from front of dunes/m | Number of plants per $m^2$ | | |
|---|---|---|---|
| | *Elymus* | *Ammophila* | *Festuca* |
| 0 | 2 | 0 | 0 |
| 5 | 4 | 0 | 0 |
| 10 | 22 | 0 | 0 |
| 15 | 29 | 0 | 0 |
| 20 | 5 | 3 | 0 |
| 25 | 6 | 6 | 0 |
| 30 | 0 | 38 | 0 |
| 35 | 0 | 17 | 0 |
| 40 | 0 | 13 | 0 |
| 45 | 0 | 8 | 16 |
| 50 | 0 | 0 | 47 |
| 55 | 0 | 0 | 91 |
| 60 | 0 | 0 | 103 |

| Abiotic factor | Distance from front of dunes/m | | |
|---|---|---|---|
| | 0 | 25 | 50 |
| Soil pH | 8.1 | 8.1 | 8.0 |
| Soil organic matter/% | 0.4 | 0.5 | 0.6 |
| Soil moisture/% | 0.3 | 0.4 | 0.4 |

**1 a Suggest a suitable graph that the student could draw to represent these data. (AO2)** 4 marks

**b The student found the soil moisture content by weighing the soil sample and then putting it in an oven at 100°C for 24 hours. She then re-weighed it. Explain how she would use these measurements to find the percentage water content of the soil. (AO2)** 2 marks

**c Name the pioneer species in the sand dune succession. (AO2)** 1 mark

**d *Ammophila* is a tall grass with long, branching roots. Use this information to suggest why *Elymus*, which is a much smaller type of grass, is not found beyond 25 m. (AO2)** 2 marks

**e Explain why the student carried out a transect in this investigation. (AO3)** 2 marks f In a different investigation, the student placed quadrats at random throughout an area. Explain how she should do this. (AO1) **2 marks**

.................................................................................................

.................................................................................................

.................................................................................................

.................................................................................................

6

## Exam-style questions

**1** **Surtsey is an island that formed in 1963 as the result of volcanic eruptions. The graph shows the number of species of plants found on the island from 1965 to 2000.**



**a Explain why:**

**i very few plants could grow on the island in the first few years after the island formed. (AO2)** **1 mark**

.................................................................................................

.................................................................................................

**ii once a few plants were growing, the number of plants that was able to grow there increased rapidly. (AO2)** **2 marks**

.................................................................................................

.................................................................................................

.................................................................................................

.................................................................................................

**iii The number of animal species found on the island also increased with a similar pattern. Explain why. (AO2)** **2 marks**

.................................................................................................

.................................................................................................

.................................................................................................

.................................................................................................

.................................................................................................

# Section 8

## The control of gene expression

Cells regulate when and how rapidly transcription and translation of the genome occurs. This regulates the cell's metabolic activity. Every cell has a copy of the whole of the organism's genome, but each specialised cell only translates a certain proportion of the genes. There are many different factors that control the expression of genes, some of which are internal factors and others that are environmental factors. Scientists now realise that the ways in which these factors interact with the genome is more complex than was once thought. Epigenetic control of gene expression is recognised as increasingly important. As humans learn to control the expression of genes, this could lead to many medical and technological advances. DNA technology can be used in the diagnosis of human diseases, and in their treatment.

### Alteration of the sequence of bases in DNA can alter the structure of proteins

Gene mutations are changes in the base sequence of DNA. These include addition, deletion, substitution, inversion, duplication and translocation of bases. They occur at random, although mutagenic agents can increase the rate of mutation. Mutations usually result in a different sequence of amino acids in the protein coded for. Some mutations affect only one triplet code. This may change the amino acid coded for, or it may have no effect. Other gene mutations change the base triplets from the point of mutation onwards. This type of mutation is called a 'frame shift'.

**1 Give *two* examples of mutagenic agents. (AO1)** 2 marks

..........

..........

**2 The base sequence of part of one strand of DNA is shown below.**

**ATTCGCGAT**

**The table below shows the same sequence after a mutation has taken place. Identify the type of mutation in each case. (AO1)** 3 marks

| Base sequence of DNA | Type of mutation |
|---|---|
| ATTGCCGAT | |
| ATTCCGAT | |
| GCGATATTC | |
| ATCTCGCGAT | |
| ATACGCGAT | |
| ATTCGCGCGAT | |

**3** **The base sequence TAGCGAATG changed to TATCGAATG after a mutation. Explain *two* possible effects of this mutation on the protein coded for. (AO1)** 3 marks

..............................................................................................................................

..............................................................................................................................

..............................................................................................................................

..............................................................................................................................

..............................................................................................................................

..............................................................................................................................

**4** **The base sequence TAGCGAATG changed to TAGGCGAATG after a mutation. Explain the effect that this change is likely to have on the protein coded for. (AO1)** 3 marks

..............................................................................................................................

..............................................................................................................................

..............................................................................................................................

..............................................................................................................................

..............................................................................................................................

..............................................................................................................................

## Gene expression is controlled by a number of features

### Most of a cell's DNA is not translated

Totipotent cells can mature into any type of body cell. They occur only for a short period of time in mammalian embryos, before becoming specialised. As cells become specialised, only some of their DNA is translated. Pluripotent stem cells can divide in unlimited numbers to become most kinds of mature cell type. These can be used to treat certain conditions in humans. Multipotent and unipotent cells also occur in mature mammals.

It is possible to form induced pluripotent stem cells (iPS cells) from unipotent cells using transcription factors. These iPS cells can potentially be used to treat disorders in humans. There are ethical and other issues involved in using stem cells to treat disease.

**1** **What is a totipotent cell and where is it found? (AO1)** 2 marks

..............................................................................................................................

..............................................................................................................................

..............................................................................................................................

..............................................................................................................................

**2** **What is a pluripotent cell and where is it found? (AO1)** 2 marks

..........

..........

..........

..........

**3** **What is the difference between multipotent and unipotent cells? (AO1)** 2 marks

..........

..........

..........

..........

**4** **What is an induced pluripotent stem cell and how is it formed? (AO1)** 3 marks

..........

..........

..........

..........

..........

..........

**5** **Scientists have discovered that about 1% of cardiac muscle cells are replaced each year. They have identified cells called cardiomyocytes that can divide to form specialised cardiac muscle cells. These cardiomyocytes are smaller than normal cardiac muscle cells, and similar to cardiomyocytes that are found in newborn babies.**

**a** **Give *two* ways in which cardiomyocytes are different from cardiac muscle cells. (AO2)** 2 marks

..........

..........

..........

..........

**b** **How might induced pluripotent stem cells be used to repair damage to the heart that is caused by a heart attack? (AO2)** 2 marks

..........

..........

..........

..........

**6** **Scientists have generated induced pluripotent stem cells from skin cells from people with neurological disorders such as Down syndrome or Parkinson's disease. They have used these induced pluripotent stem cells to generate brain cells to help us to understand the causes of these disorders and to develop new drugs to treat them.**

a **How can scientists turn specialised skin cells into induced pluripotent stem cells? (AO1)** 1 mark

..................................................................................................................................................................

..................................................................................................................................................................

b **How is the ability to generate brain cells helping scientists to develop new drugs for neurological disorders? (AO2)** 2 marks

..................................................................................................................................................................

..................................................................................................................................................................

..................................................................................................................................................................

..................................................................................................................................................................

## Regulation of transcription and translation

In eukaryotes, specific transcriptional factors move from the cytoplasm into the nucleus to stimulate or inhibit transcription of specific genes. Hormones such as oestrogen can stimulate transcription of certain genes.

Epigenetics describes inherited changes in gene function that do not involve a change to the DNA base sequence. These changes are triggered by environmental factors and are important in the development and treatment of diseases such as cancer.

Another way in which gene expression is controlled is by RNA interference.

**7** **Describe how oestrogen can stimulate the transcription of certain genes. (AO1)** 4 marks

..................................................................................................................................................................

..................................................................................................................................................................

..................................................................................................................................................................

..................................................................................................................................................................

..................................................................................................................................................................

..................................................................................................................................................................

..................................................................................................................................................................

..................................................................................................................................................................

**8** **What is 'epigenetics'? (AO1)** 2 marks

..................................................................................................................................................................

..................................................................................................................................................................

..................................................................................................................................................................

..................................................................................................................................................................

**9** What effect do the following have on gene expression?

a Increased methylation of the DNA. (AO1) **1 mark**

........................................................................................................................

........................................................................................................................

b Decreased acetylation of histone proteins associated with the DNA. (AO1) **1 mark**

........................................................................................................................

........................................................................................................................

**10** Scientists have recently studied the genomes of a newborn baby and a 103-year-old man. They found that there were significantly lower levels of DNA methylation in a subset of analysed genes in the old man, when compared with the newborn. What does this suggest? (AO2) **2 marks**

........................................................................................................................

........................................................................................................................

........................................................................................................................

........................................................................................................................

**11** Scientists suspect that epigenetic changes may be factors in psychiatric disorders, diabetes and cancer, as well as in ageing. Scientists have examined the DNA of 86 sets of twin sisters aged 30 years to 80 years. They found that 490 genes linked with ageing showed signs of methylation. The younger sets of twins displayed lower levels of methylation in a range of genes when compared with the older sets of twins. Methylation can be triggered by lifestyle factors such as smoking and environmental stresses.

a What do these findings suggest about methylation and ageing? (AO2) **2 marks**

........................................................................................................................

........................................................................................................................

........................................................................................................................

........................................................................................................................

b Suggest why the scientists were using sets of identical twins in their investigation. (AO2) **3 marks**

........................................................................................................................

........................................................................................................................

........................................................................................................................

........................................................................................................................

........................................................................................................................

........................................................................................................................

**12** **Put the following statements in order to describe how RNA interference works. (AO1)** 2 marks

......................................................................................................................................

A **This separates the two strands into the passenger and guide strand.**

B **This siRNA binds to an RNA-induced silencing complex (RISC).**

C **dsRNA in the cell's cytoplasm is cut by an enzyme called Dicer into double-stranded small interfering RNA (siRNA) molecules, which are 20–25 nucleotides long.**

D **This cuts the mRNA so that the unwanted target protein is not produced and the gene is 'silenced'.**

E **The passenger strand is degraded, while the RISC takes the guide strand to a specific mRNA site.**

......................................................................................................................................

## Gene expression and cancer

A tumour is a group of rapidly dividing cells. Benign tumours tend to stay in one place, but malignant tumours have cells that break off and spread to other parts of the body. This is cancer. Normal cell division is regulated by tumour suppressor genes and proto-oncogenes. When these genes mutate, or become abnormally methylated, cancer may develop. Some breast cancers are associated with increased oestrogen concentration.

**1** **Describe how the following regulate cell division: (AO1)** 2 marks

a **proto-oncogenes**

......................................................................................................................................

......................................................................................................................................

......................................................................................................................................

......................................................................................................................................

b **tumour suppressor genes**

......................................................................................................................................

......................................................................................................................................

......................................................................................................................................

......................................................................................................................................

**2** **Describe how mutations in proto-oncogenes and tumour suppressor genes can lead to cancer. (AO1)** 3 marks

......................................................................................................................................

......................................................................................................................................

......................................................................................................................................

......................................................................................................................................

......................................................................................................................................

......................................................................................................................................

**3** **Describe the difference between benign and malignant tumours. (AO1)** **4 marks**

........................................................................................................................

........................................................................................................................

........................................................................................................................

........................................................................................................................

........................................................................................................................

........................................................................................................................

........................................................................................................................

........................................................................................................................

**4** **Describe how methylation of a tumour suppressor gene might lead to cancer. (AO1)** **2 marks**

........................................................................................................................

........................................................................................................................

........................................................................................................................

........................................................................................................................

**5** **The graph shows the incidence of colon cancer in women, and the daily meat consumption per person in many different countries. Each point on the graph represents a different country.**

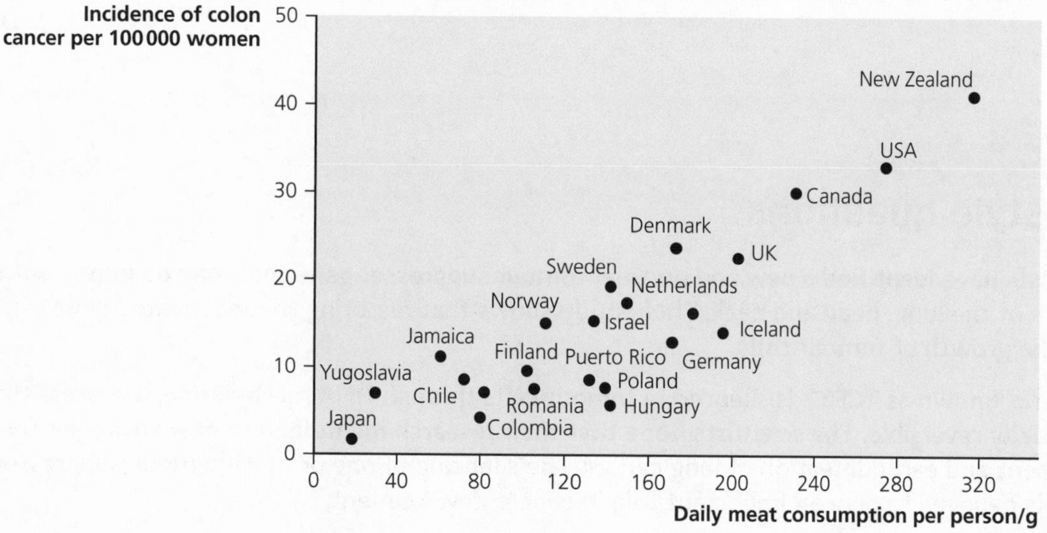

**a** **Describe the graph. (AO2)** **2 marks**

........................................................................................................................

........................................................................................................................

........................................................................................................................

........................................................................................................................

b A journalist saw this graph and wrote an article titled 'Eating meat causes colon cancer'. Do you agree with this conclusion? Explain your answer. (AO3) **4 marks**

................................................................................................................................

................................................................................................................................

................................................................................................................................

................................................................................................................................

................................................................................................................................

................................................................................................................................

................................................................................................................................

................................................................................................................................

**6** Some kinds of breast cancer are stimulated by oestrogen. Tamoxifen is a drug that is used to treat this kind of breast cancer. Tamoxifen binds to oestrogen receptors. Explain how this is useful in treating breast cancer. (AO2) **3 marks**

................................................................................................................................

................................................................................................................................

................................................................................................................................

................................................................................................................................

................................................................................................................................

................................................................................................................................

10

## Exam-style questions

**1** Scientists have identified a new and unusual tumour suppressor gene that may be important in cancers of the lung, head and neck. Their study shows that restoring the inactivated gene can slow the growth of tumour cells.

The gene, known as TCF21, is silenced in tumour cells through DNA methylation, a process that is
05 potentially reversible. The scientists hope that their research might lead to new strategies for the treatment and early detection of lung cancer. The silencing of one or more tumour-suppressor genes is believed to play an important role in cancer development.

The newly discovered gene is unusual because it can change a normal epithelial cell into a more primitive kind of cell that is usually found in embryos, and which is capable of migrating to
10 other tissues. This suggests that silencing of the TCF21 gene might help a tumour to spread to other areas of the body, a process known as metastasis. It has also been shown that this gene is commonly silenced or lost in a variety of other cancers, including breast and ovarian cancer, melanoma and lymphoma. This makes the scientists think that the gene could be important in many forms of cancer.

a What is a tumour suppressor gene (line 1)? **2 marks**

................................................................................................................................

................................................................................................................................

................................................................................................................................

................................................................................................................................

b **Describe how DNA methylation can silence TCF21 (line 4).** 2 marks

.......................................................................................................

.......................................................................................................

.......................................................................................................

.......................................................................................................

c **Cancer is more likely to be treated successfully if it is detected earlier (line 6). Explain why.** 2 marks

.......................................................................................................

.......................................................................................................

.......................................................................................................

.......................................................................................................

d **Scientists believe that if they could restore the activity of TCF21, this would be important in treating some kinds of cancer. Use the information in the passage and your own knowledge to explain why.** 3 marks

.......................................................................................................

.......................................................................................................

.......................................................................................................

.......................................................................................................

.......................................................................................................

.......................................................................................................

## Using genome projects

The ability to find the base sequence of stretches of DNA has become faster and more efficient through the development of new sequencing methods that are continuously being improved and updated. These methods have allowed scientists to sequence the genomes of many organisms, including humans. Sequencing the genomes of simpler organisms has allowed scientists to increase their understanding of these organisms' proteomes. There are many applications for this knowledge and understanding, including the development of new vaccines. In higher organisms it is harder to work out the proteome because of the presence of non-coding DNA and regulatory genes.

1 **Distinguish between the *genome* and the *proteome*. (AO1)** 2 marks

.......................................................................................................

.......................................................................................................

.......................................................................................................

2 **Explain how understanding the genomes of simple organisms can lead to an understanding of their proteomes. (AO1)** 3 marks

.......................................................................................................

.......................................................................................................

.......................................................................................................

.......................................................................................................

**3** Why is it harder for scientists to work out the proteome of a higher organism than the proteome of a simpler organism? (AO1) **2 marks**

................................................................................................................................

................................................................................................................................

................................................................................................................................

................................................................................................................................

**4** Scientists have sequenced the genome of the parasite that causes malaria. They are hoping to develop a vaccine against malaria, but it is difficult to do this because the parasite keeps changing its antigens.

a What is an antigen? (AO1) **2 marks**

................................................................................................................................

................................................................................................................................

................................................................................................................................

................................................................................................................................

b Use your knowledge of natural selection to explain why the parasite keeps changing its antigens. (AO2) **4 marks**

................................................................................................................................

................................................................................................................................

................................................................................................................................

................................................................................................................................

................................................................................................................................

................................................................................................................................

................................................................................................................................

c Recently, scientists have found that the malaria parasite invades red blood cells by binding to a receptor on the cell called 'basigin'. The parasite has a surface protein called RH5 that binds to basigin. The gene that codes for RH5 is very similar in all malaria parasites. Suggest an explanation for this. (AO2) **4 marks**

................................................................................................................................

................................................................................................................................

................................................................................................................................

................................................................................................................................

................................................................................................................................

................................................................................................................................

................................................................................................................................

d The scientists think that the RH5 protein would be a good antigen to use as the basis for a vaccine. Explain why. (AO2) **3 marks**

................................................................................................................................

................................................................................................................................

................................................................................................................................

................................................................................................................................

## Gene technologies

Sections of DNA may be transferred from one organism to another. This works because DNA is a universal code, and is transcribed and translated in the same way in all living organisms. The organism that receives the new DNA is called a transgenic organism. There are several ways of obtaining the required gene. The gene can be copied many times by in vivo or in vitro methods. Promoter and terminator regions may be added to the genes. The DNA can then be inserted into a vector using restriction endonuclease and ligase enzymes. Vectors are then used to transform host cells. Marker genes may be used to identify transformed cells.

Gene therapy is a related technique that may be useful in medicine. There are many ethical, financial and social issues connected with the use of DNA technology.

**1** **Complete the table to show the type of enzyme that carries out each function. (AO1)** 4 marks

| Enzyme | Function |
|---|---|
| | Cuts DNA at a specific base sequence |
| | Makes a single-stranded piece of complementary DNA from a messenger RNA template |
| | Joins two pieces of DNA together |
| | Joins DNA nucleotides together to make a piece of DNA |

**2** **DNA is described as a 'universal code'. Explain what this means. (AO1)** 2 marks

..........

..........

..........

..........

**3** **A scientist wanted to synthesise the gene coding for insulin using a 'gene machine'. Describe the stages in this process. (AO1)** 3 marks

..........

..........

..........

..........

..........

**4** **The insulin gene could also be synthesised using mRNA from pancreatic cells. Describe how. (AO1)** 2 marks

..........

..........

..........

..........

**5** **What are 'sticky ends' and how are they useful in gene technology? (AO1)** 3 marks

..........

..........

..........

..........

..........

**6** **Annotate the diagram to show how the bacterium *E. coli* can be genetically engineered to produce a human protein such as insulin. (AO1)** 4 marks

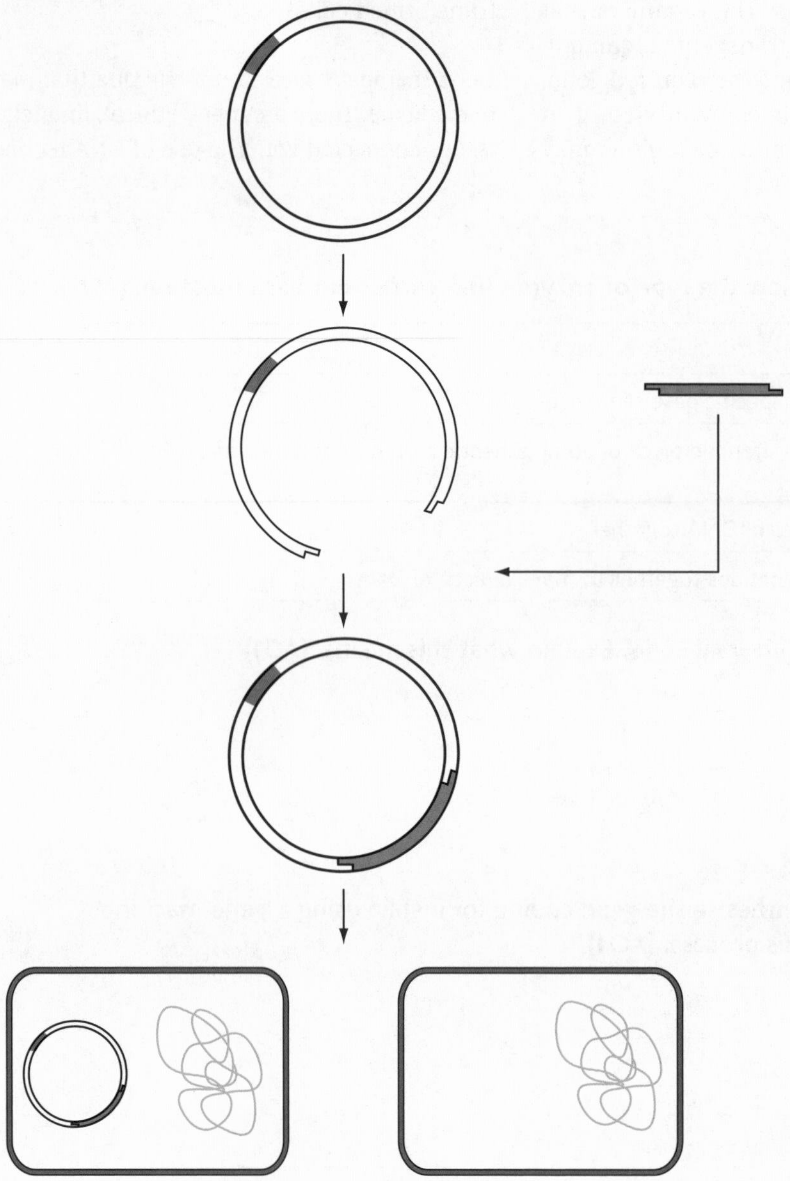

**7** **What is a marker gene and what is it used for? (AO1)** 3 marks

..........................................................................................

..........................................................................................

..........................................................................................

..........................................................................................

**8** **What is a *vector* in genetic engineering? Give *two* examples. (AO1)** 2 marks

..........................................................................................

..........................................................................................

..........................................................................................

..........................................................................................

**9** **Describe an in vivo method of cloning a gene or a section of DNA. (AO1)** 3 marks

..........

..........

..........

..........

..........

..........

**10** **A scientist used the polymerase chain reaction to make many copies of a piece of target DNA. He put the target DNA into a tube, along with the other ingredients needed. Name the other ingredients needed. (AO1)** 3 marks

..........

..........

..........

..........

**11** **What is a *primer*? Explain why two primers are needed for PCR. (AO1)** 2 marks

..........

..........

..........

..........

**12** **The diagram summarises the stages in PCR.**

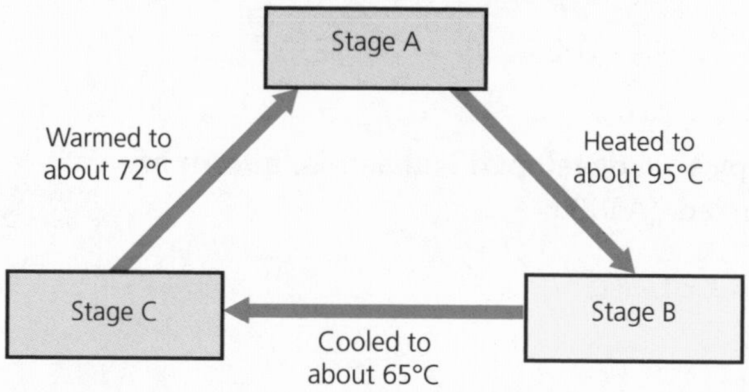

**Describe what happens at:**

**a** **stage B (AO1)** 2 marks

..........

..........

**b** **stage C (AO1)** 2 marks

..........

..........

**c** **stage A (AO1)** 2 marks

..........

..........

**13** What is gene therapy? (AO1) **2 marks**

..............................................................................................................................

..............................................................................................................................

..............................................................................................................................

..............................................................................................................................

**14** Some children are born with a serious condition called severe combined immunodeficiency (SCID). It is caused by their white blood cells being unable to make an enzyme called ADA. As a result, the immune systems of these children do not work properly, and they are susceptible to many infections.

In the early twenty-first century, the following gene therapy trial was carried out:

- Doctors removed multipotent stem cells from the bone marrow of several children with SCID.
- The stem cells were incubated with a modified virus that contained the gene for ADA.
- The transformed cells were returned to the bone marrow of the children.

Several of the children were cured of SCID as a result, and developed a functional immune system. Some of the other children had to repeat the procedure at a later date.

**a** What was the role of the virus in this trial? (AO1) **1 mark**

..............................................................................................................................

..............................................................................................................................

**b** Suggest why some of the children had to repeat the procedure later while others did not. (AO2) **2 marks**

..............................................................................................................................

..............................................................................................................................

..............................................................................................................................

**c i** Three of the children in this gene therapy trial developed leukaemia, a form of cancer, and died. Suggest how this occurred. (AO2) **2 marks**

..............................................................................................................................

..............................................................................................................................

..............................................................................................................................

..............................................................................................................................

**ii** Evaluate whether the doctors were right to carry out this gene therapy trial. (AO3) **4 marks**

..............................................................................................................................

..............................................................................................................................

..............................................................................................................................

..............................................................................................................................

..............................................................................................................................

..............................................................................................................................

..............................................................................................................................

..............................................................................................................................

## Differences in DNA between individuals of the same species can be exploited for the identification and diagnosis of heritable conditions

Specific alleles can be detected in DNA samples using DNA probes that hybridise with complementary DNA sequences. This means that people can be screened for specific alleles. The information can be used in genetic counselling, and in personalised medicine.

**15 What is a gene probe? (AO1)** 2 marks

..............................................................................................................................................

..............................................................................................................................................

..............................................................................................................................................

**16 Why must a gene probe be radioactive or fluorescent? (AO1)** 2 marks

..............................................................................................................................................

..............................................................................................................................................

..............................................................................................................................................

**17 The diagram shows the appearance of a gel after DNA has been separated by gel electrophoresis.**

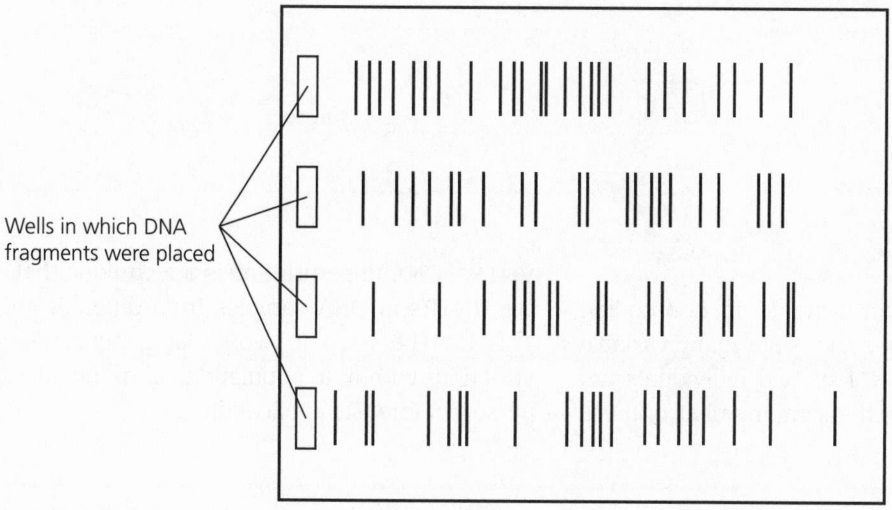

Wells in which DNA fragments were placed

**a Pieces of DNA move towards the positive electrode. Explain why. (AO2)** 2 marks

..............................................................................................................................................

..............................................................................................................................................

..............................................................................................................................................

..............................................................................................................................................

**b Draw a circle around the shortest piece of DNA. (AO1)** 1 mark

**c The DNA on the gel is treated with alkali to make it single stranded before a probe is applied. Explain why. (AO1)** 1 mark

..............................................................................................................................................

..............................................................................................................................................

**18** **The BRCA1 and BRCA2 alleles greatly increase the chance that a woman will develop breast cancer at some point in her life. It is possible to screen a person to see if they carry these alleles. If they do, they will be offered a radical mastectomy (removal of breast tissue), which makes it very unlikely that they will develop breast cancer.**

**A young woman of 18 has a grandmother who died of breast cancer. Her mother is being treated for breast cancer and has tested positive for both the BRCA1 and BRCA2 alleles. Doctors have offered this young woman the opportunity to be screened for these alleles. What factors should she consider when reaching her decision? (AO3)** 4 marks

..............................................................................................................................

..............................................................................................................................

..............................................................................................................................

..............................................................................................................................

..............................................................................................................................

..............................................................................................................................

**19** **Explain how personalised medicine and genome sequencing can be useful in the treatment of diseases such as cancer. (AO1)** 2 marks

..............................................................................................................................

..............................................................................................................................

..............................................................................................................................

..............................................................................................................................

## Genetic fingerprinting

A great deal of the human genome is non-coding DNA. Within the non-coding regions are many variable number tandem repeats (VNTRs). Two individuals are very unlikely to have exactly the same number of these VNTRs. DNA fingerprinting is a technique that analyses the VNTRs in DNA samples from different individuals. This can be used to work out genetic relationships, variation within a population, and the identity of a person in forensic applications.

**20** **What are VNTRs? (AO1)** 2 marks

..............................................................................................................................

..............................................................................................................................

..............................................................................................................................

**21** **A person has a different number of VNTRs at a particular locus on one chromosome from the number of VNTRs at the same locus on their homologous chromosome. Explain why. (AO2)** 2 marks

..............................................................................................................................

..............................................................................................................................

..............................................................................................................................

..............................................................................................................................

22 A woman has just had a baby. She claims that a wealthy premier league football player is the father of the child, but he denies this. Explain how genetic fingerprinting could be used to settle this argument. (AO1) 3 marks

23 Describe how genetic fingerprinting is carried out. (AO1) 6 marks

24 Suggest how a zoo could use genetic fingerprinting to establish the genetic variability of chimpanzees in a captive population. (AO2) 3 marks

25 A woman whose brother has cystic fibrosis wants to know whether she is a carrier of the CF allele. Describe how gel electrophoresis and a gene probe could be used to screen for this allele. (AO1) 5 marks

# Also available

Go to http://www.hoddereducation.co.uk/studentworkbooks for details of all our student workbooks.


Hodder Education, an Hachette UK company, Blenheim Court, George Street, Banbury, Oxfordshire OX16 5BH

*Orders*

Bookpoint Ltd, 130 Milton Drive, Milton Park, Abingdon, Oxfordshire OX14 4SE

tel: 01235 827827
fax: 01235 400401
e-mail: education@bookpoint.co.uk

Lines are open 9.00 a.m.–5.00 p.m., Monday to Saturday, with a 24-hour message answering service. You can also order through www.hoddereducation.co.uk





ISBN 978-1-4718-4502-4

First printed 2016

Impression number 5 4 3 2 1
Year 2020 2019 2018 2017 2016

HODDER EDUCATION
LEARN MORE





Printed in Spain

Hachette UK's policy is to use papers that are natural, renewable and recyclable products and made from wood grown in sustainable forests. The logging and manufacturing processes are expected to conform to the environmental regulations of the country of origin.